ENERGY and FUELS

Troll Associates

ENERGY and FUELS

by Laurence Santrey

Illustrated by Ray Burns

Troll Associates

Library of Congress Cataloging in Publication Data

Santrey, Laurence.
    Energy and fuels.

    Summary: Explains how energy used for various
activities is released by burning wood, coal, oil,
gas, and other fuels, as well as by the wind, moving
water, nuclear fission, and the sun.
    1. Power resources—Juvenile literature. 2. Fuel
—Juvenile literature. [1. Power resources. 2. Fuel]
I. Burns, Raymond, 1924-   ill. II. Title.
TJ163.23.S26   1985   333.79   84-2704
ISBN 0-8167-0290-X (lib. bdg.)
ISBN 0-8167-0291-8 (pbk.)

Copyright © 1985 by Troll Associates, Mahwah, New Jersey.
All rights reserved. No part of this book may be used or
reproduced in any manner whatsoever without written permission
from the publisher.
Printed in the United States of America.

10 9 8 7 6 5 4 3 2 1

It takes energy to run a car, to power an electric drill, and to boil water on a stove. The energy for these activities comes from fuels being burned. There are many different kinds of fuels that release energy when they are burned. Among these fuels are wood, coal, oil, and gas. Energy is also released by the wind, moving water, nuclear fission, and the sun.

In one way or another, most of the energy on Earth comes from the sun. Without the sun, plants would not grow. And if there were no plants, there would be no wood and no fossil fuels.

Fossil fuels, such as coal and oil, are fossilized remains of plants and animals that lived many millions of years ago. Time, heat, and pressure changed them into liquids, solids, and gases that can be burned to release energy.

For many thousands of years, firewood was the primary fuel used by people. It was burned in a fireplace, on an open hearth, or in a stove. It provided energy for cooking and for warmth in homes. Wood fueled the fires that blacksmiths needed to soften metal for horseshoes, nails, and other items. Wood also heated the kilns in which pottery was made.

Before the Industrial Revolution and the invention of power-driven machinery, relatively little fuel was used, because most goods were made by hand. Then, in the eighteenth and nineteenth centuries, people learned how to harness energy to drive machines. These power-driven machines could do much more physical work than any human being. But it took a great deal of fuel to run machines.

In the steam engine, the heat from burning fuel turned water into steam. The steam drove the pistons that turned the wheels that pushed a steamboat through the water or a train along a track. Steam engines also powered machines in factories and mills. Industry and large-scale production, in the modern sense, had begun.

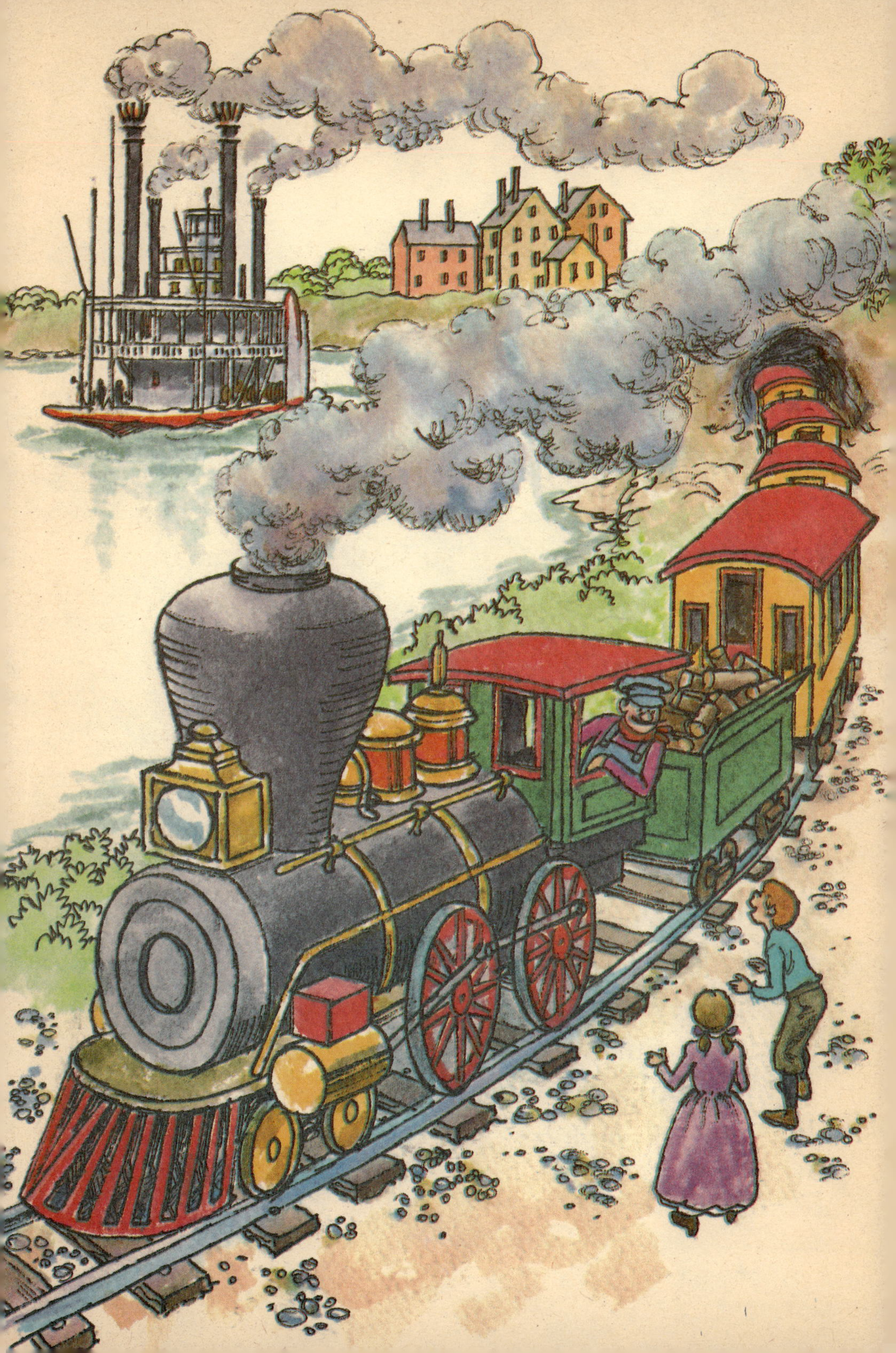

During the Industrial Revolution the demand for energy grew tremendously. People began to look for other fuels besides wood to drive machines, and better and more efficient ways to use those fuels. It was at this time that coal became the major fuel used in industrial countries. It remained so until oil replaced coal as the number-one fuel.

Coal, like wood, is a solid fuel. Coal is made of fossilized ferns, mosses, trees, and other plants that lived millions of years ago. As these plants died, they piled up in thick layers at the bottom of the forests. As more plants grew and died, they piled up on top of them. This process continued for a very long time.

Eventually, the dead plant matter turned into a spongy, brown substance called peat. Peat can be used as fuel, but it does not burn very efficiently. In some countries, however, where there is little coal and oil, peat is used as a home fuel.

Sometimes shifts in the Earth's surface buried beds of peat, and immense heat and pressure built up. The great heat and pressure changed the peat into lignite, or brown coal. Lignite burns more efficiently than peat, because it contains less moisture.

When lignite undergoes even greater underground pressure and heat over a long period of time, it is changed into coal. There are two forms of coal, bituminous and anthracite.

Bituminous

Anthracite

Bituminous, or soft, coal is used extensively in industry. It is burned in furnaces for the energy it produces and is also the source of chemicals, plastics, coke, and gas fuel.

One of the problems in using bituminous coal is that it gives off a considerable amount of smoke, ash, and other air pollutants. For example, bituminous coal contains high levels of sulfur dioxide. When this kind of coal is burned, it gives off sulfur and nitrogen dioxide. These chemicals mix with moisture in the air to form acids.

Acids formed this way are carried down to Earth when the moisture falls down as rain. This acid rain, which may reach the Earth's surface many miles from where the coal was burned, can change lake and river waters so that no fish can live in them. It can also kill plant life and damage buildings, bridges, and roads.

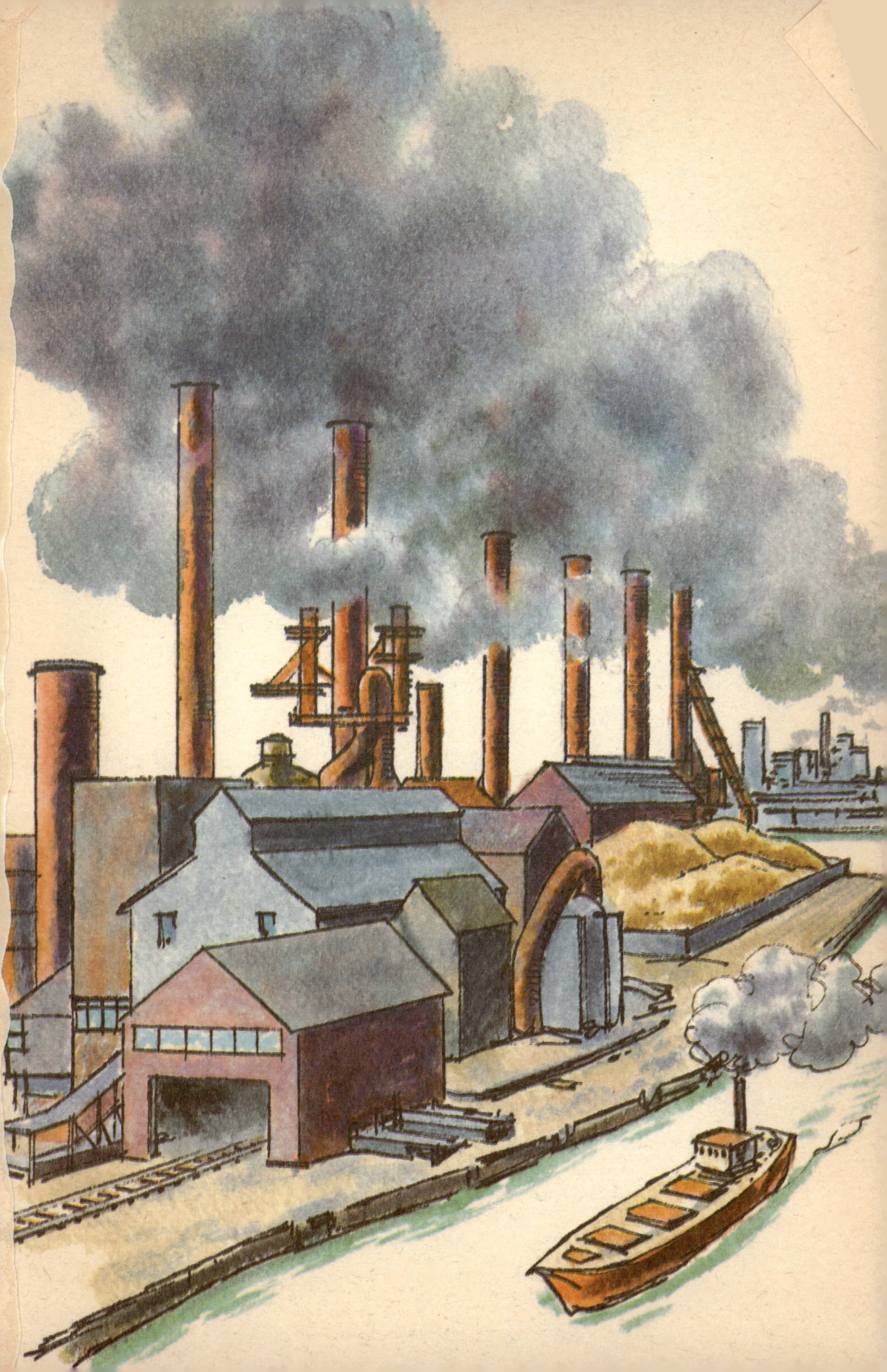

Some of the pollutants released by burning soft coal can be removed by devices called scrubbers. These scrubbers are installed in the exhaust stacks through which the fumes rise. Pollutants can also be reduced by treating soft coal with certain chemicals before burning it. These methods, however, do not remove all the pollutants.

Anthracite, or hard coal, burns more evenly and more efficiently than soft coal. It also burns with relatively little smoke and ash. But we have much less anthracite coal than bituminous coal. About one-third of the world's energy needs are supplied by coal—but only a small fraction of that coal is anthracite.

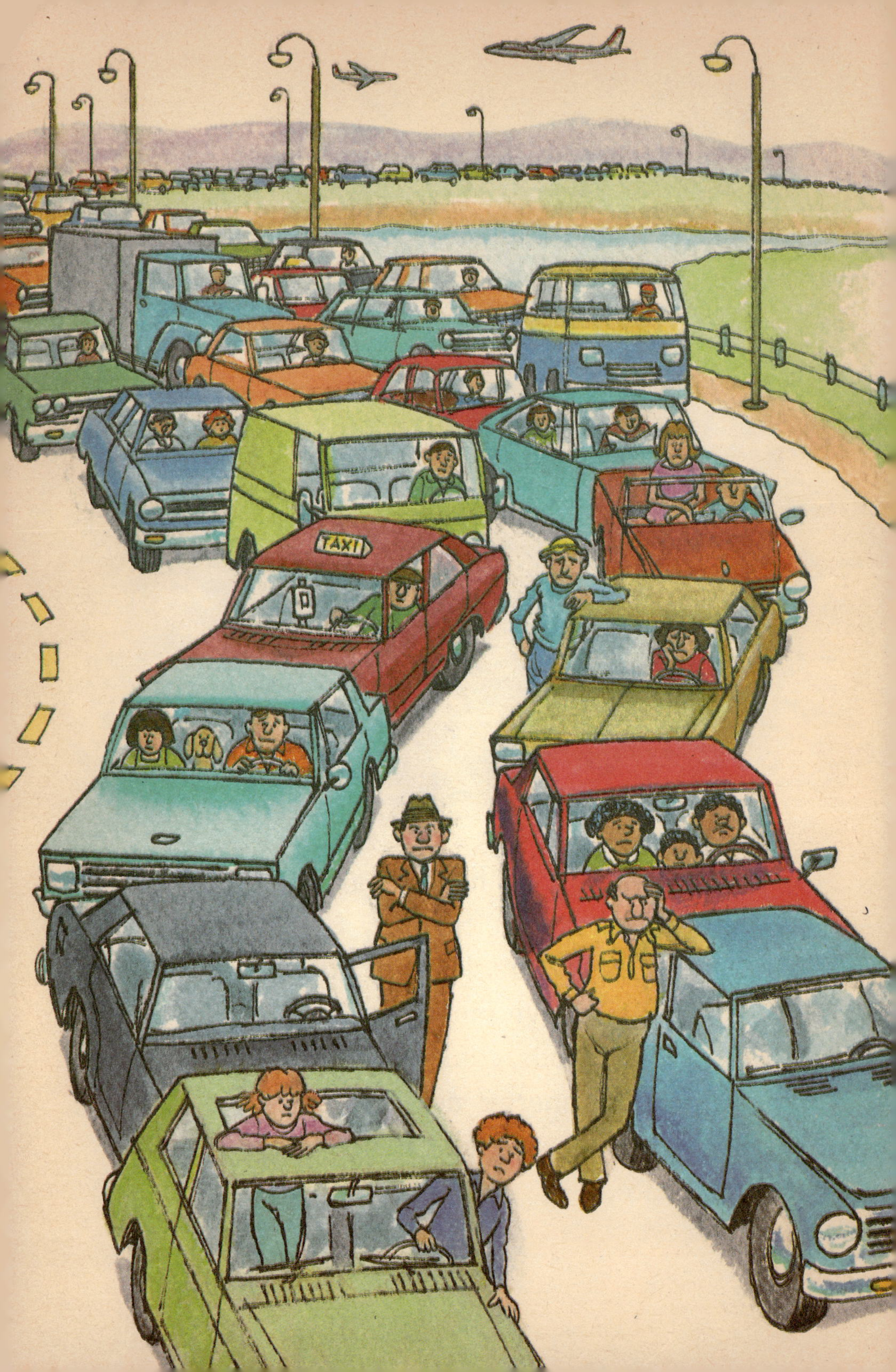
TAXI

The most widely used fuel in the world is petroleum. Liquid fuels, such as petroleum, were not very important fuels until the middle of the nineteenth century. It was then that the internal-combustion engine was developed. Since that time, the use of petroleum has increased until—at present—almost half of all the fuel used on Earth is petroleum.

In the ground, petroleum is a thick liquid sometimes called crude oil. In this form, petroleum does not burn very efficiently. To become useful fuel, the crude oil is refined, separated, and treated. Refining produces kerosene, gasoline, diesel fuel, heating oil, and lubricating oils.

For many years petroleum products were inexpensive and readily available for industry and transportation. There was a rapid growth in the use of petroleum products, particularly in the United States. Today, however, we realize that the Earth's supply of oil is limited, and that we must conserve it.

Natural gas is usually found in the ground near petroleum beds and probably was formed at the same time. Natural gas is the cleanest fossil fuel available. It causes far less pollution than oil or coal. Although there are vast reserves of natural gas below the surface of the Earth, the supply is not endless. Eventually the supplies of this fuel will be exhausted.

Nuclear fuels are a recent addition to the world's energy sources. In nuclear *fission*, uranium atoms are split inside a machine called a reactor. As the atoms split, enormous amounts of energy are released. This energy can be used to produce electricity in a power plant, or be used to power ships, submarines, and other craft.

In nuclear *fusion*, atoms are combined, or fused, and energy is released. But so much energy is released that scientists have not been able to control the nuclear fusion process. Perhaps in the future, nuclear fuels will safely provide for all of our energy needs.

First, however, science must solve many problems. One problem is the disposal of atomic waste materials so that they will not poison the Earth and its inhabitants.

Another danger is that if nuclear reactions ever got out of control, they could cause widespread destruction. Until science makes nuclear energy truly safe, the world must continue to rely on other sources of energy.

The machines so essential to modern life may be powered in the future by a wide variety of energy sources. These include solar power, wind power, fuel cells, geothermal energy, solid-waste treatment, hydrogen, the ocean tides, and other sources not yet imagined.

Solar panel

Nuclear-fusion reactor

Windmill

Fuel cell

Geothermal well

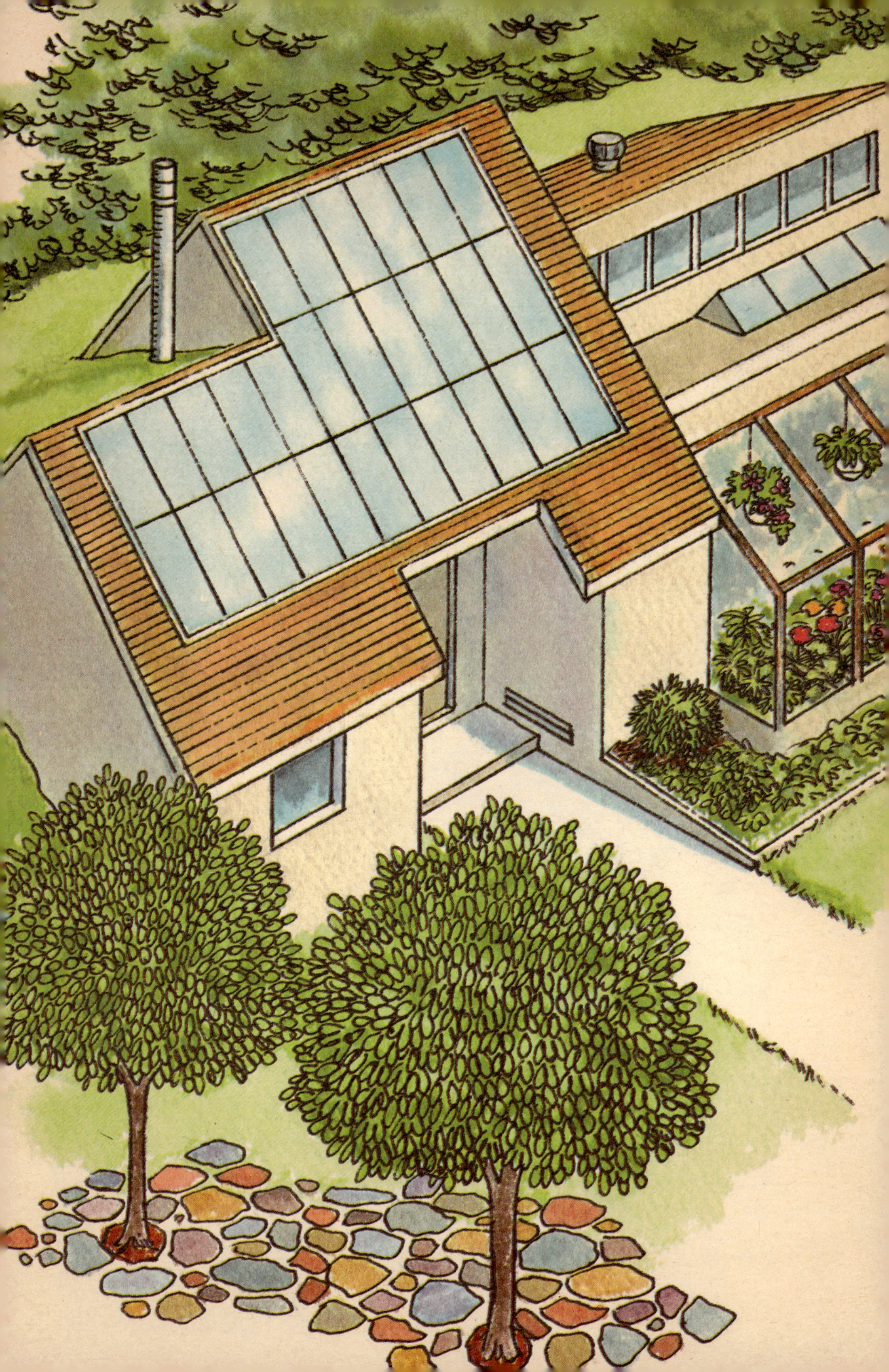

Solar energy, of course, is already being used in many ways today. Solar cells are used to convert light into electricity, and solar collectors are used to gather the sun's heat. Solar energy is pollution free, and its supply—the sun—is practically endless.

Unfortunately, the sun does not shine evenly over the Earth. So, until technology makes it possible to store and transmit solar energy, this energy form can only supplement our other fuels.

Two other energy sources—wind and tides—are also not evenly distributed over the Earth. So, like solar energy, wind and tidal energy are still limited solutions to the world's energy problems.

Future technology may make the fuel cell a key answer to our energy needs. Fuel cells, which mix hydrogen and oxygen, produce energy for spacecraft such as the American space shuttle. But fuel cells are very expensive at the present time. Not until they are less costly can they be considered for general use.

Scientists know how to use hydrogen— one of the elements of water—as a fuel. They know how to harness geothermal energy, or heat that exists deep below the Earth's surface. They know how to remove oil from underground shale and how to make synthetic fuels. But all these processes are so expensive that they are not yet practical.

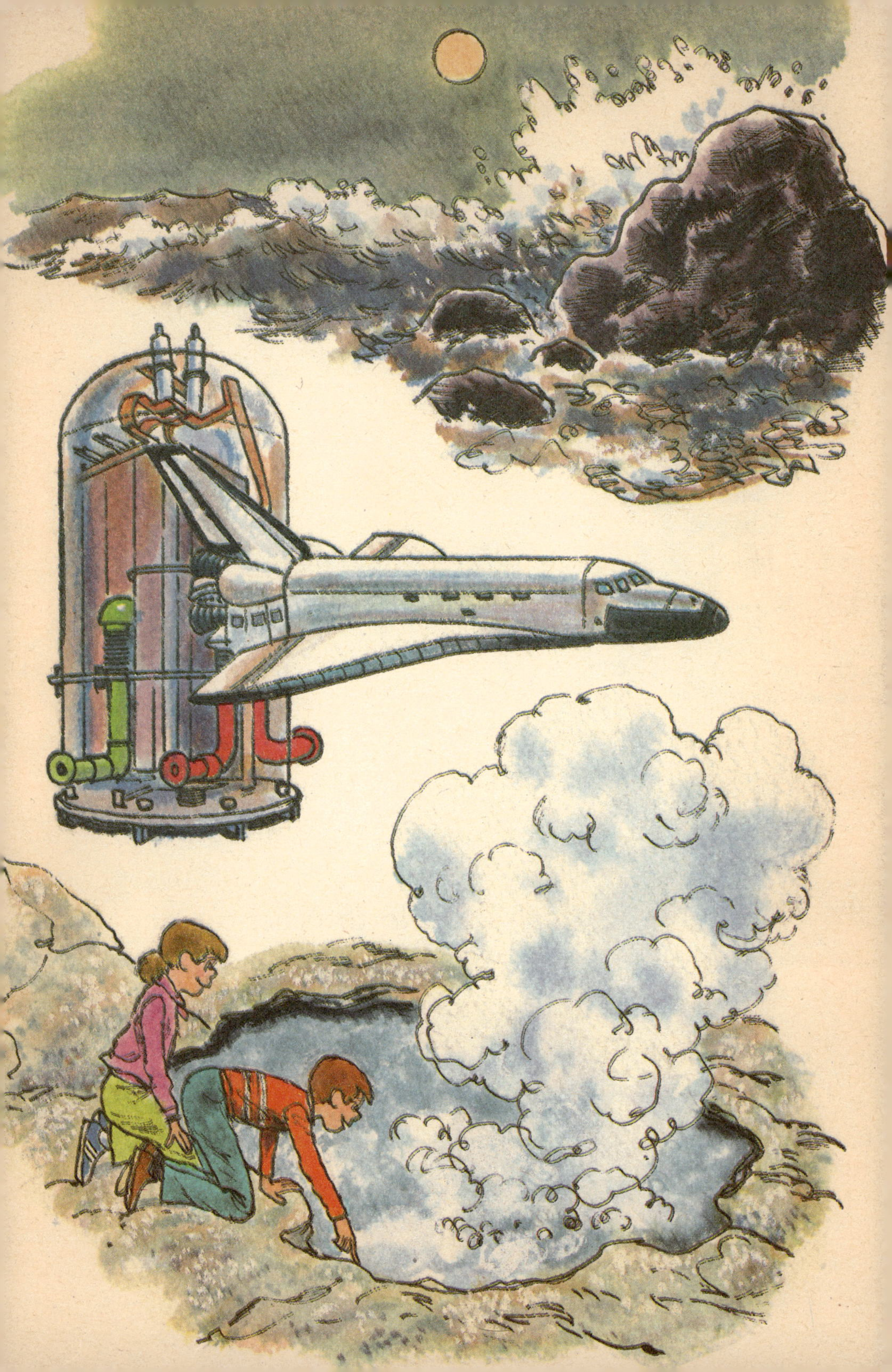

Until we have cheap, nonpolluting, unlimited fuels at our disposal, we must use our fossil fuels wisely.

In recent years, the North American and European countries, which use most of the world's energy, have reduced their consumption of fossil fuels.

People everywhere hope that conservation will continue even while science works to bring us new forms of energy.